FROM **SEA** TO **SALT**

by Robin Nelson

Lerner Publications Company / Minneapolis

Lerner Publications Company
A division of Lerner Publishing Group, Inc.
241 First Avenue North
Minneapolis, MN 55401 U.S.A.

Website address: www.lernerbooks.com

Library of Congress Cataloging-in-Publication Data

Nelson, Robin, 1971-
    From sea to salt / by Robin Nelson.
      p. cm. — (Start to finish)
    Summary: Briefly introduces the process by which table salt is made from sea salt.
    ISBN-13: 978-0-8225-0946-2 (lib. bdg. : alk. paper)
    ISBN-10: 0-8225-0946-6 (lib. bdg. : alk. paper)
    1. Salt industry and trade—Juvenile literature. 2. Seawater—Juvenile literature. [1. Salt industry and trade.] I. Title. II. Start to finish (Minneapolis, Minn.)
TN905.N45 2004
664'.4   2002152928

Manufactured in the United States of America
2 3 4 5 6 7 – DP – 13 12 11 10 09 08

The photographs in this book appear courtesy of: © Todd Strand/Independent Picture Service, cover, p. 23; © C. Lujan/Photo Network, pp. 1 (top), 5; © AP Photo, *The Wichita Eagle*, Brian Corn, pp. 1 (bottom), 21; © Phyllis Picardi/Photo Network, p. 3; © Photri/W. Kulik, p. 7; © Cargill Salt, pp. 9, 11, 13, 17; © Carolina Biological/ Visuals Unlimited, p. 15; © AP Photo, *Salt Lake Tribune*, Leah Hogsten, p. 19.

Table of Contents

I like salt on food.

Where does salt come from?

Seawater dries up.

Salt comes from the sea. Seawater covers most of the earth. Long ago, there was even more seawater. Some of it dried up. It left behind land. It left behind salt, too.

Workers find salt.

Workers find the salt by digging a hole called a well. Most wells are very deep.

7

Water is sprayed on the salt.

The salt is very hard. Workers spray water on the salt. The water breaks the salt into tiny pieces. Together, water and salt are called **brine**. A pipe moves the brine out of the well.

BRINE

SETON NAME PLATE CORP. 800-243-6624

The brine is moved to a tank.

Machines pump the brine to a building called a **plant**. The plant has many machines. The brine is put into a large tank.

The brine is heated.

The brine is moved to a machine. The machine heats the brine to dry up the water.

Solid salt pieces are made.

Soon most of the water dries up. The salt turns into solid pieces called **salt crystals**. The salt crystals and the water that are left behind are called **slurry**. This picture shows slurry very close up.

The slurry is dried.

The slurry is moved to a machine called a dryer. This dryer is not for clothes. The dryer removes the rest of the water from the slurry.

The salt is screened.

The dry salt is moved to a **screen**. A screen is a piece of metal with small holes in it. The screen separates big salt crystals from small salt crystals. We use small salt crystals on our food.

The salt is put into boxes.

A machine puts the salt into boxes or bags. Most salt for people to eat is put into boxes. Trucks take the salt to stores for people to buy.

Salt adds flavor to my food.

People like to put salt on many different foods. I like salt on potato chips and vegetables. Salt can help food taste good!

Glossary

brine (BRYN): tiny pieces of salt in water

plant (PLANT): a building with many machines

salt crystals (SAHLT KRIH-stuhlz): solid salt pieces

screen (SKREEN): a piece of metal with tiny holes

slurry (SLUR-ee): solid salt pieces and water

Index